Karakul Sheep

Farmer's Bulletin No. 1632

by US Dept. of Agriculture

with an introduction by Jackson Chambers

.

Self Reliance Books

Get more historic titles on animal and stock breeding, gardening and old fashioned skills by visiting us at:

Introduction

I am pleased to present yet another practical title on breeding and raising livestock.

The work is in the Public Domain and is re-printed here in accordance with Federal Laws.

As with all reprinted books of this age that are intended to perfectly reproduce the original edition, considerable pains and effort had to be undertaken to correct fading and sometimes outright damage to existing proofs of this title. At times, this task is quite monumental, requiring an almost total "rebuilding" of some pages from digital proofs of multiple copies. Despite this, imperfections still sometimes exist in the final proof and may detract from the visual appearance of the text.

I hope you enjoy reading this book as much as I enjoyed making it available to readers again.

Jackson Chambers

KARAKUL SHEEP are important chiefly because they produce lambskins suitable for fur. The lambskins are divided into three general classes designated by the trade names broadtail, Persian, and caracul.

The hardiness of this breed, which originated in Bokhara, in west-central Asia, would be a valuable characteristic if bred into native flocks in some sections of the Southwestern States where the climate is similar to that of Bokhara.

The purebred Karakul sheep in the United States are not numerous. Only three importations from their native country have been made, one in 1909, one in 1913, and the last in 1914. Further direct importations have been prohibited by law in order to prevent the introduction into the United States of animal diseases prevalent in Asia.

Karakul sheep now on hand can be used for grading up by mating selected purebred Karakul rams with ewes of the more available American types. Experimental work has indicated that high-grade Karakul sheep that will be as satisfactory as purebred Karakuls in the production of commercial fur can be developed by mating successive generations of select-grade ewes with purebred Karakul rams that are especially prepotent in transmitting desirable fur characteristics.

Washington, D. C.

Issued August 1930
Revised October 1938

KARAKUL SHEEP

By C. G. Potts, *animal husbandman*, and V. L. Simmons, *assistant animal husbandman, Animal Husbandry Division, Bureau of Animal Industry*

CONTENTS

BREED CHARACTERISTICS

THE KARAKUL BREED of sheep is of the fat-tail type, the tail being broad, flat, and tapering rapidly toward the end. The fat stored in the tail enables the sheep to withstand long periods of scant feed. The head is peculiarly characteristic of the breed, the face being black and narrow, the skull much rounded, with Roman nose and thin, black, pendulous ears. The legs, as well as the face and ears, are covered with short, fine, lustrous, black hair. The ewes are generally hornless but the rams usually have long, spiral, outspreading horns. (Fig. 1.)

The wool is coarse and long and varies in color from black through various shades of brown to light gray. Practically all lambs are black at birth and continue that color until they are yearlings, after which the wool gradually becomes a lighter shade. However, this change of color is not uniform in all animals, some remaining rather dark in maturity, whereas others become very light. The title-page illustration shows a purebred Karakul ewe and lamb. Note the tight curl of the lamb's coat, which has a high luster. This is the type of skin desired for fur.

The breed is noted for its hardiness and is able to exist and do well under very adverse conditions. In their native home, Bokhara, in west-central Asia, the sheep are kept in the mountains through the summer and are driven to the lowlands for wintering when the mountains are covered with snow. Although the Karakul is found in its greatest purity in Bokhara, especially in the Karakul Valley, animals with a large percentage of Karakul blood are found in various other countries as far west as the Caspian Sea, and also in that section of the southern part of the Union of Soviet Socialist Republics adjacent to the Black and Caspian Seas.

IMPORTATIONS OF KARAKUL SHEEP

Three importations of Karakul sheep from Bokhara, totaling 34 rams and 33 ewes, were brought into the United States in 1909, 1913, and 1914. The offspring of these sheep are now widely scattered throughout the United States and Canada, the greatest numbers of both grades and purebreds in the United States being in Texas, California, Colorado, New York, Idaho, Wisconsin, Michigan, and Minnesota.

CLASSES OF LAMBSKINS

This breed is particularly prized for the production of lambskins, utilized as fur, which may be roughly divided into three classes, commercially known as broadtail, Persian, and caracul. Of these three classes the broadtail is the most valuable, but the percentage of this type of skin produced is very small. It is produced usually by lambs prematurely born, although cases are reported of this type of

FIGURE 1.—Good type of Karakul ram. The coarse, wavy, lustrous fleece is typical of the best individuals of the breed.

fur being produced by lambs born at full fetal development. It is a flat, lightweight fur with a water-wave pattern which is very lustrous and beautiful. The fact that practically all lambs from which this most valuable type of lamb fur is taken would be a total loss (since they are stillborn) in the production of other breeds is one of the important advantages in producing the Karakul breed. Broadtail skins are used almost entirely for making ladies' coats. (Fig. 2.)

"Persian lamb" is a type of skin produced by lambs of the Karakul breed which are from 3 to 10 days old. It is necessary to watch the skins carefully, because the curl opens rapidly after the fifth day, and while the value increases with the size as long as the curl remains tight, it is essential that the pelt be removed before the quality of the fur is reduced. In some cases this is as late as the tenth day after birth. As far as is known, no artificial means has been developed of increasing the curl or in any way changing the pattern after birth. This type of fur (fig. 3) is used in making ladies' coats and in trimming coats of other materials for both men and women.

"Caracul" is a trade name given to a lustrous, open type of fur (fig. 4), which shows a wavy or moiré pattern and is entirely free from close curls. These skins are usually light in weight and are very desirable as materials for ladies' coats. They occur naturally in black as well as in various shades of brown and tan. The tendency

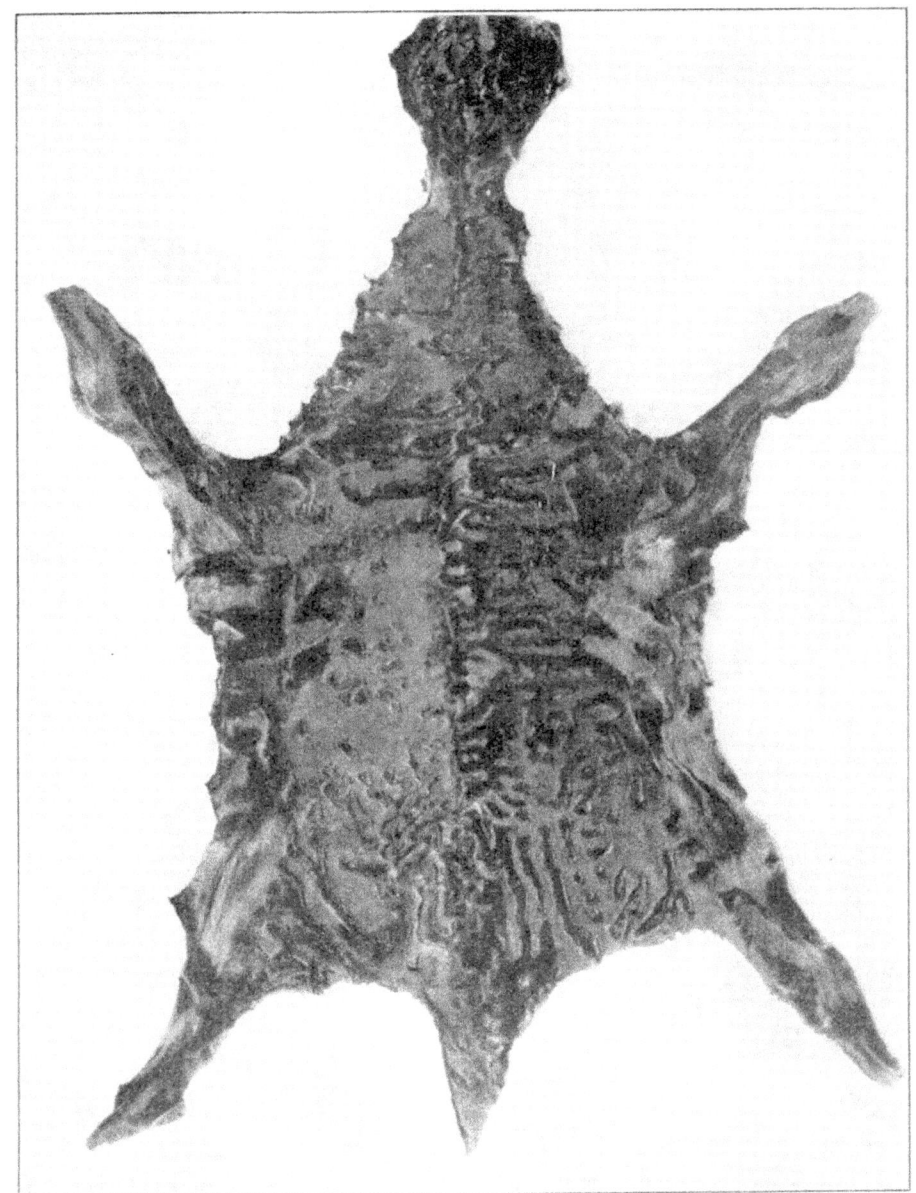

FIGURE 2,—Skins classed as "broadtail." These skins should be extremely lustrous and of light weight.

to lighter shades is an individual characteristic of the type of sheep producing this kind of fur. The skins are best when removed at an age not greater than 2 weeks, but they do not deteriorate with age so rapidly as either the broadtail or the Persian type. They can be used for fur as long as the luster is retained, which may be as long as 2 months. The term "caracul" means this type of fur and not the Karakul breed of sheep, in spite of the similarity of the names and the fact that caracul fur is produced by lambs of the Karakul breed.

Considerable variation in type and quality of lambskin fur is found in all Karakul flocks, whether purebred or grade. While occasional skins of exceptional quality occur, the average must be greatly improved to equal the average quality of imported skins. Karakul skins

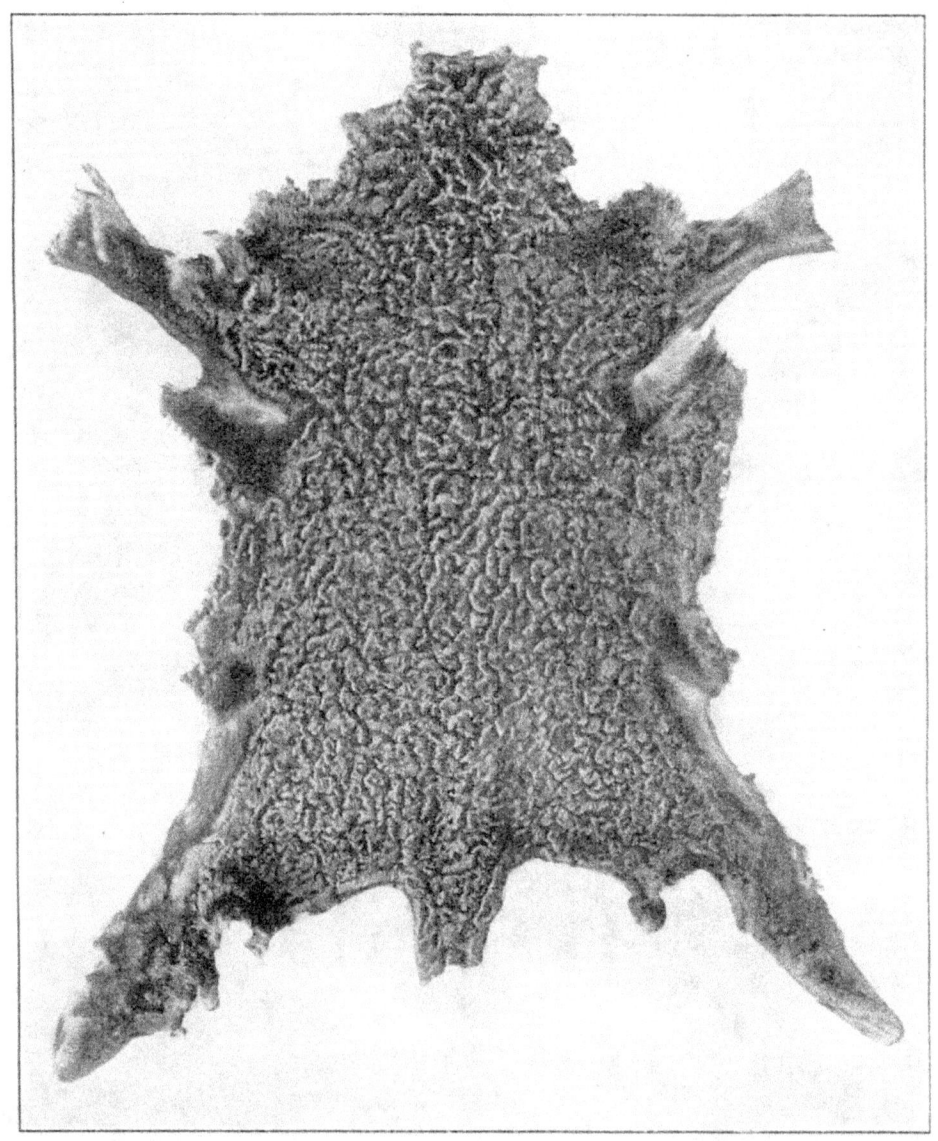

FIGURE 3.—Persian lambskin. This type of fur is more open, with looser curls, than broadtail. Its value is determined largely by uniformity of pattern and lustrous appearance.

produced in the United States that sell at prices ranging from $12 upward are extremely rare.

Various methods have been used in curing the raw skins for shipment. Some are dried in the shade, some pickled, and others are treated with a mixture of barley meal and salt while the fur is sprinkled with flour or sand to keep the curl from opening up. They should never be doubled or salted, as this may cause the skin to break. In packing, the fur sides of the skins should always be placed together to avoid injury by the contact of the raw skin with the fur.

KARAKUL WOOL

The wool that is sheared from Karakul sheep is lustrous but coarse and varies in color from a light gray through various shades of brown to black. It ranges in length from 8 to 12 inches, and the fleece weights usually range from 7 to 9 pounds. This wool can be made into beautiful rugs and carpets and is also used in the manufacture of blankets and automobile robes. However, its uses are rather limited when com-

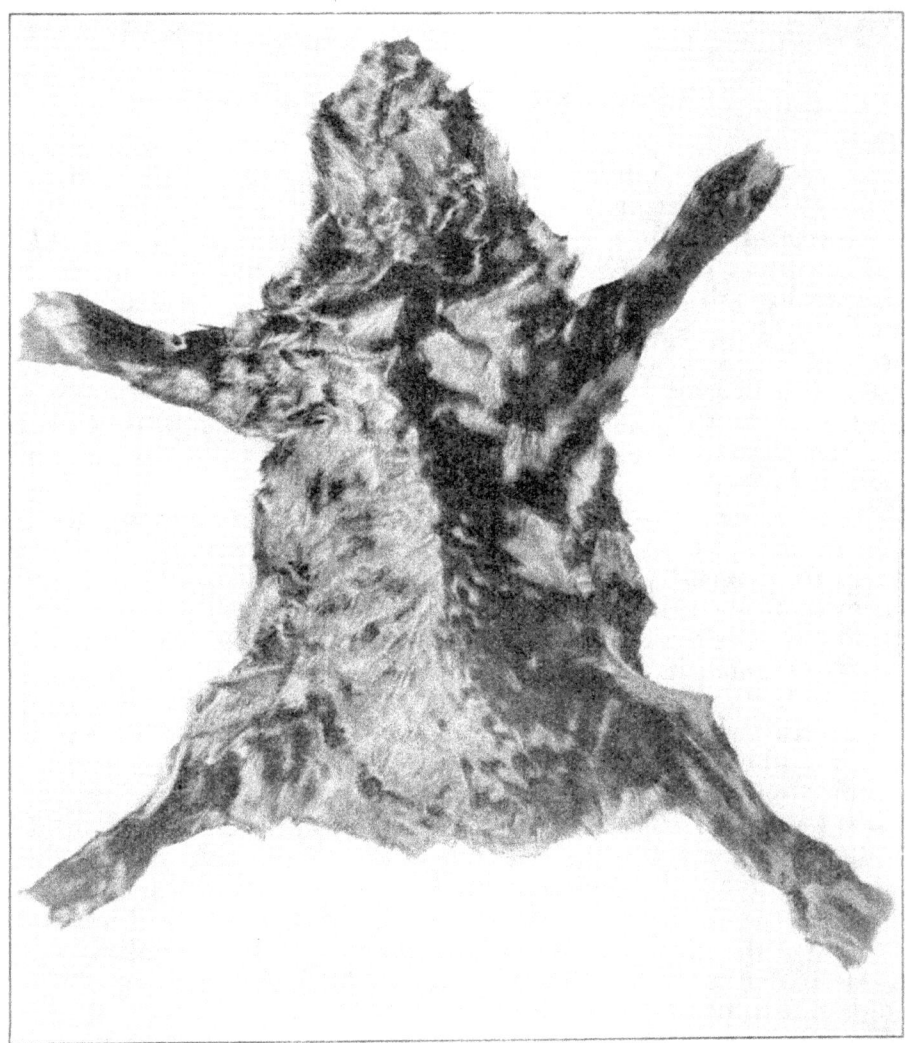

FIGURE 4.—Skin classed as "caracul." The value of this type of fur depends on the character of the wave and the luster it shows. The most valuable skins are light weight, with a pronounced, satiny sheen.

pared with those of the finer grades, and when handled through regular marketing channels Karakul wool sells at from one-half to three-fourths the price received for good-quality white wool produced by the breeds of sheep that are more thoroughly established in the United States.

CHARACTERISTICS OF THE MEAT

While Karakul lambs are hardy and grow out well, they are rather angular in conformation and are not so well developed in those portions of the carcass that yield the most expensive cuts of meat as are the more specialized mutton breeds. The hardiness of the lambs is due, no doubt, to the ability of the breed to withstand the long periods of short feed and the semiarid conditions that prevail so extensively in the country in which they are commonly raised.

Since the Karakul is primarily a fur-producing breed, no attempt has ever been made to develop a mutton strain. Nevertheless, the meat is wholesome, and the wether lambs can be marketed to defray part of the expense of production while grading up the ewe flock to a point where the pelts of the lambs will be more valuable as fur than the lambs would be for meat if grown to marketable size.

UTILIZATION OF THE BREEDING STOCK

Only a small number of Karakul sheep have been imported into the United States, and further direct importations from the countries of which the breed is native have been prohibited by legal restrictions on account of certain serious diseases of animals in those countries. Multiplication of the Karakul sheep already on hand in this country must be through breeding, and their qualities may be extended further by grading up other breeds to the Karakul type by the use of purebred Karakul rams. Purebred Karakul ewes, of course, should be mated with purebred Karakul rams. This purity of breeding can be maintained, and at the same time a ram bred to a limited extent to purebred Karakul ewes may also be mated with a rather large number of ewes of other more available American types.

Five successive crosses of the female offspring to purebred Karakul rams result in sheep that are thirty-one thirty-seconds, or approximately 97 percent, pure Karakul breeding. But it is not expected that ewes of even this high grade will be as valuable for the production of fur as is the best imported stock. Since, however, some of the imported animals have not proved to be satisfactory for the production of valuable fur, it is doubtful whether imported Karakul sheep, on which little or no definite information can be obtained before shipment, would be better than selected grade animals of a high percentage of choice, pure breeding, or even as valuable.

Karakul sheep have been crossed to a limited extent with Rambouillet, Merino, Lincoln, Cotswold, and other breeds in the United States, and good, marketable skins have been produced by this grading-up process. Though information on the results is limited, the true Karakul type is evidently prepotent in the best individuals of known parentage, which trace in pure blood lines through the sire and dam to some of the best imported animals.

The hardiness of Karakul sheep is a valuable characteristic, which may be introduced, through breeding, into some of the common American breeds. Karakul inheritance in flocks of grade sheep in some dry or semiarid regions of the United States where the climate is similar to that of Bokhara doubtless would enable the grade sheep to withstand adverse weather conditions better than the native flocks.

CROSS-BREEDING AND GRADING UP

The Bureau of Animal Industry has conducted cross-breeding experiments in which Karakul rams have been mated with Cotswold, Cheviot, Merino, Barbados, Corriedale, and Blackfaced Highland ewes.

Of eight skins taken from lambs produced by Cotswold ewes bred to Karakul rams only three were valued as high as $1 each in the raw state, although all were black and showed considerable curl, as do the skins of practically all first-cross lambs sired by Karakul rams. A skin may be black and curly and yet have little value, because it lacks luster and has a poor type of curl.

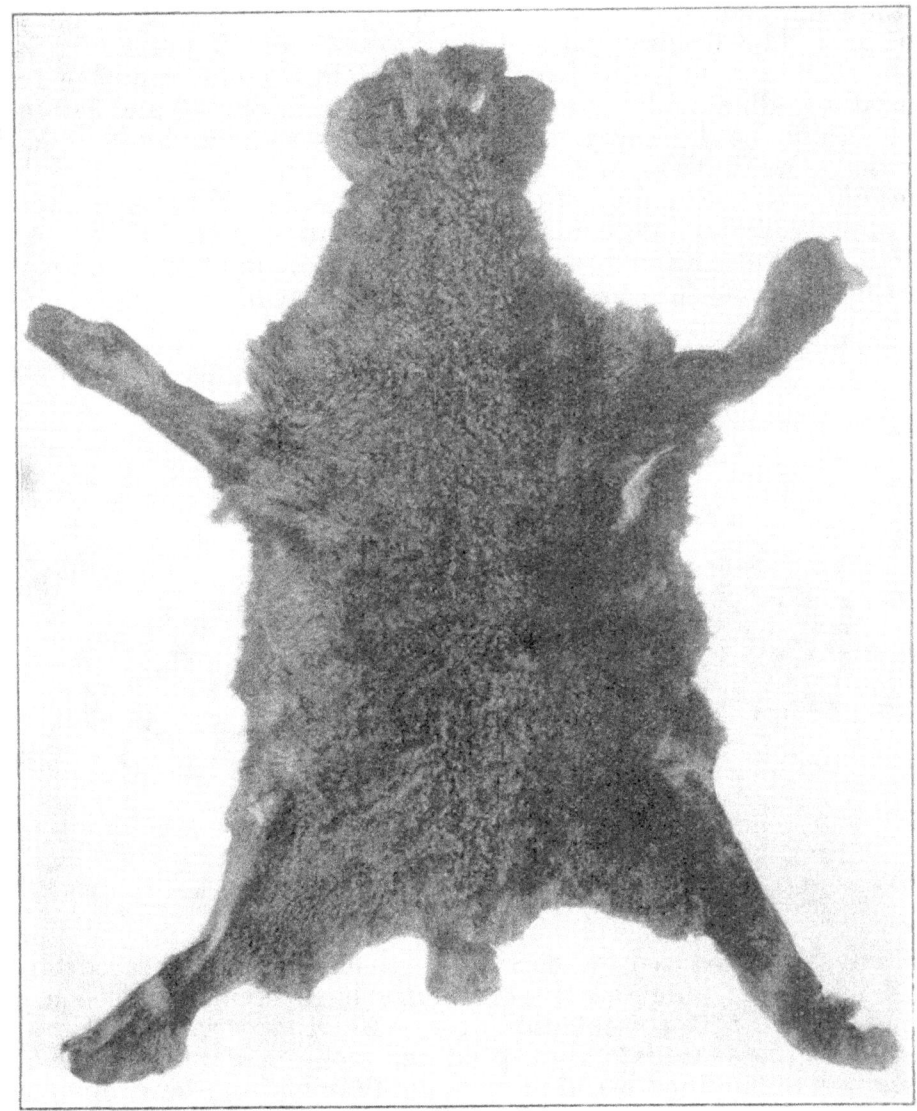

FIGURE 5.—Skin from first-cross Karakul-Merino lamb—black but poor in character of curl and luster. This skin has no fur value.

Six skins were procured from lambs produced by Cheviot ewes bred to Karakul rams. One skin was valued at $3 and one at $1. The others did not have sufficient value as fur to pay the charge for "dressing," the trade term for preparing the skins for use as fur.

Of five skins from lambs of Merino ewes sired by Karakul rams, none had sufficient fur value to repay the charge of dressing. These skins were particularly poor in luster, and the character of curl was still poorer than in the other crosses mentioned above (fig. 5).

Fifty-nine lambs were produced from the mating of a Karakul ram with Barbados ewes. Barbados are a woolless type of sheep that have a short, rather stiff, hairy coat, and it was thought that these might be suitable for crossing with the Karakul rams. The skins of the 59 first-cross lambs had very little fur value (fig. 6). Fifteen of the first-cross ewe lambs were mated to a second imported Karakul ram in the spring of the year and to a third Karakul ram in the fall of the same year. The skins of the second-cross lambs resulting from the spring breeding ranged in value from $1 to $10, averaging $4.70 each. The skins obtained from the second-cross lambs resulting from the fall breeding, with the exception of two valued at $3 and $10 each, had little or no fur value. All appraisals were made by New York furriers. While seasonal climatic conditions and feed may have had some effect on the quality of the skins produced by this lot of ewes in two successive lambings, much of the variation was believed to be traceable to the greater prepotency of the sire of the first lot in transmitting desirable fur characteristics to his offspring.

FIGURE 6.—Cross-bred Karakul-Barbados ewe lambs, showing loose curls and dull fleeces, which are typical of the first-cross lambs, and are undesirable for fur.

Lambskins produced by crossing Corriedale and Blackfaced-Highland ewes with purebred Karakul rams have been of sufficient fur value to justify the continuation of a series of top crosses of Karakul rams with these breeds.[1] Ewes produced from these crosses have been bred back to purebred Karakul rams until lambs from four top crosses have been obtained for fur studies. Fur specialists of the Department have found that the lambskins from the fourth top crosses average sufficiently high in quality to be readily marketable through the regular fur-trade channels.

The average values placed on the lambskins produced in the spring of 1936 at the Agricultural Research Center, Beltsville Md., are shown in table 1.

[1] These cross-breeding experiments with Karakul sheep have been conducted cooperatively by the Bureau of Animal Industry and the Bureau of Biological Survey of the U. S. Department of Agriculture.

TABLE 1.—*Comparison of the fur value of lambskins representing from one to four top crosses of purebred Karakul rams on Corriedale and Blackfaced-Highland ewes*

Lambskins (number)	Breeding of lambs	Purity of breeding	Value of lambskins	
			Range	Average
3	Karakul × Corriedale	½ Karakul	$0.33–$0.92	$0.72
3	do	¾ Karakul	2.00– 4.25	3.47
9	do	⅞ Karakul	.83– 5.83	3.33
5	do	¹⁵⁄₁₆ Karakul	3.00– 3.58	3.33
10	Karakul × Blackfaced Highland	¾ Karakul	1.58– 5.75	3.03
8	do	⅞ Karakul	.50– 6.50	3.38
2	do	¹⁵⁄₁₆ Karakul	4.08– 7.17	5.62
29	Karakul	Purebred	.67– 7.75	4.26

The average raw-pelt values shown in table 1 were assigned by three disinterested furriers on the New York fur market in the spring of 1936. The figures show that pelts from the second top cross and up compare favorably with the purebred Karakul produced at the Research Center.

Lambskins representative of the type and quality of fur produced in the Department's flock from the different top crosses of purebred Karakul rams on Corriedale ewes are shown in figures 7, 8, 9, and 10. The improvement in the quality of the fur that may be obtained by additional top crosses is indicated by the average value placed on these dressed skins by New York furriers. These skins were priced as follows: One-half Karakul, $0.75 (fig. 7); three-fourths Karakul, $4.25 (fig. 8); seven-eighths Karakul, $5.25 (fig. 9); and fifteen-sixteenths Karakul, $6.50 (fig. 10). The skins represent approximate average values on 3 years' data.

In general, results of the Department's Karakul sheep experiments thus far indicate that lambskins suitable for fur can be produced not only with purebred Karakuls but also through continued top-crossing with purebred Karakul rams on ewes from a foundation of the breeds commonly available in the United States, namely, Cotswold, Cheviot, Merino, Corriedale, and Blackfaced Highland. From an economic point of view, profitable fur production in this country depends mainly on using selected purebred Karakul rams in grading up flocks of high-grade Karakul ewes developed from inexpensive foundation ewe stock of the established breeds available in the United States.

The need for grading up flocks of high-grade Karakuls is due to the scarcity and high prices that prevail for purebred Karakul sheep.

GRADE KARAKUL RAMS AS SIRES

The scarcity of purebred Karakul rams and the high value placed on them has led to the use of grade rams in many cases. Since a half-blood ram has a dark fleece and shows some waviness of the fleece even when mature, he will often appear to be valuable as a sire, especially to those inexperienced in this type of sheep production. Although, as stated, such rams, when bred to longwool ewes, may sire lambs that are black and curly, the skins of these lambs will not necessarily be valuable as fur.

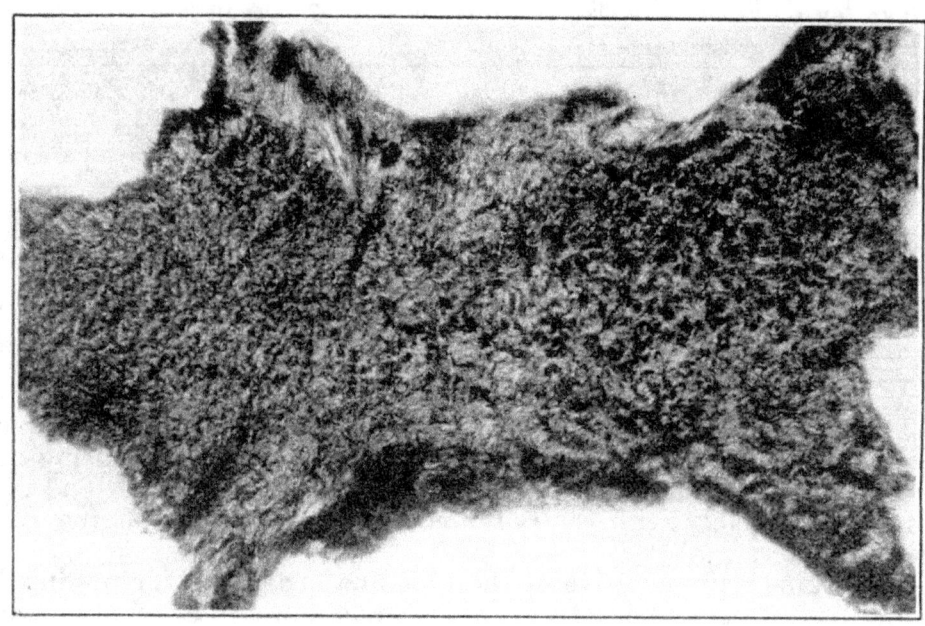

FIGURE 7.—Appearance of skin of a lamb from a first cross of a Karakul ram on a Corriedale ewe. Note the loose curl and lack of luster.

FIGURE 8.—Skin of a lamb from a second cross of a Karakul ram on a Corriedale-Karakul
ewe. This lamb is three-quarters Karakul. Note the improvement in curl and luster
over the skin shown in figure 7.

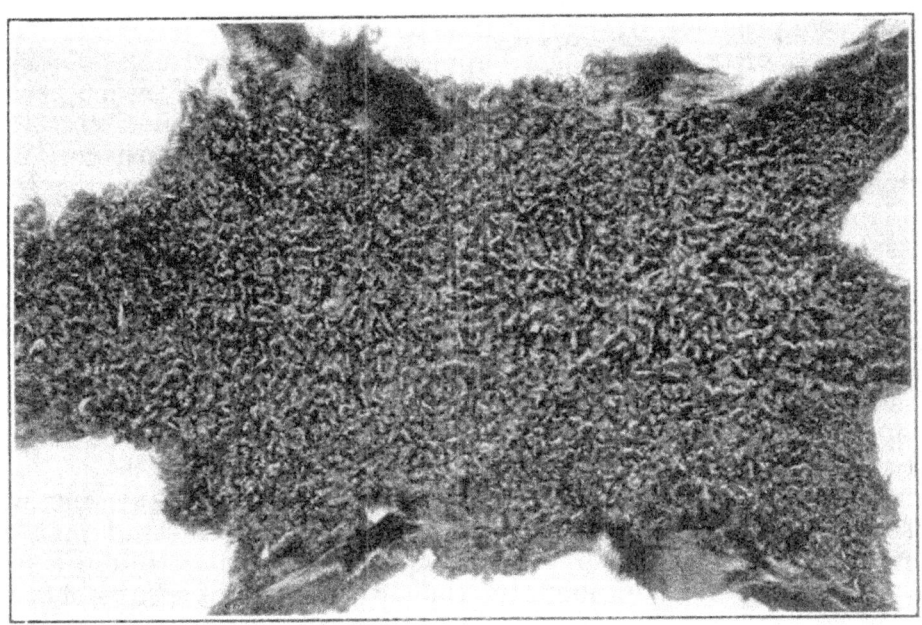

FIGURE 9.—Skin of a lamb from a third cross of a Karakul ram on a Corriedale-Karakul ewe. This lamb is seven-eighths Karakul and shows improvement in both curl and luster.

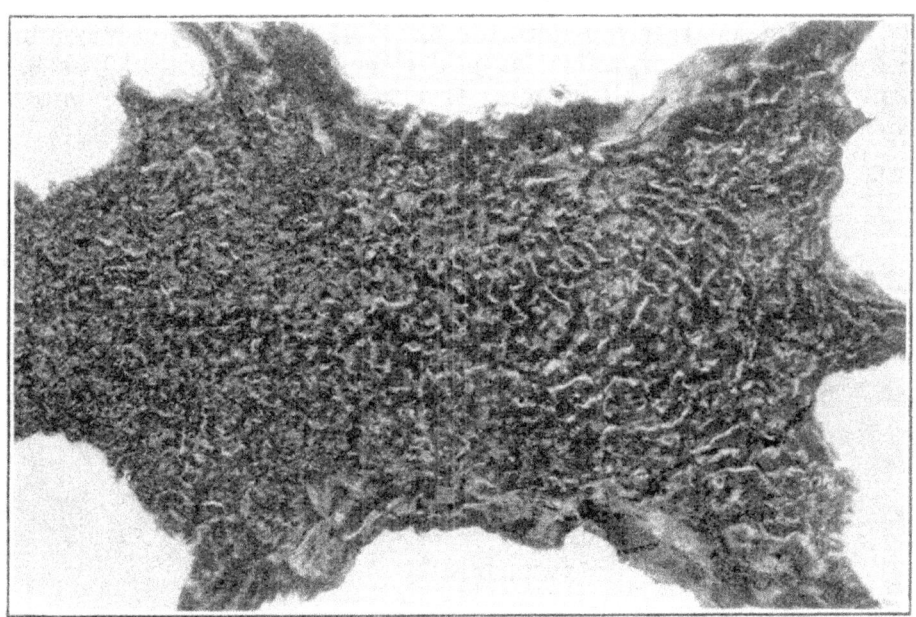

FIGURE 10.—Appearance of skin of a fifteen-sixteenths Karakul lamb. This is the fourth cross of a Karakul ram on a Corriedale-Karakul ewe.

In one experiment four lambs sired by cross-bred Karakul-Barbados rams were produced by ewes of the same breeding as the rams. These lambs were of the same general appearance as the first-cross Karakul-Barbados lambs that had no value for fur purposes. In another experiment the same spring seven Karakul-Barbados first-cross ewes were bred to a three-quarter-Karakul-one-quarter-Barbados ram, which had at birth a skin valued at $10. As a result of these matings 10 lambs were produced that theoretically were five-eighths Karakul blood, but none of them had skins of fur value.

The results of these experiments provided further evidence that little could reasonably be expected in the production of valuable fur from the interbreeding of grade Karakul rams and ewes. Some benefit from the use of these grade rams may be derived, however, through the infusion of Karakul blood into flocks of other breeds, for foundation stock to be graded up later for fur production by the use of purebred Karakul rams. While there is little or no possibility of producing marketable furs from the mating of grade Karakul rams with grade Karakul ewes, the use of selected purebred Karakul rams for this purpose shows much promise of success. The results of these tests and the personal knowledge the authors have of results obtained by breeders, in grading up flocks carrying a preponderance of Karakul blood, indicate that a type of sheep can be developed that will be entirely satisfactory for the production of marketable skins. Success in developing such a type of sheep will depend on the continued use of the best purebred Karakul rams now available in this country on successive generations of choice, grade Karakul ewes.

The various organizations in the United States that have been maintaining separate registers for Karakul sheep have now combined to form a national organization under the name "Karakul Fur Sheep Registry," in which all animals tracing in all ancestry to imported stock may be recorded.

ORGANIZATION OF THE UNITED STATES DEPARTMENT OF AGRICULTURE WHEN THIS PUBLICATION WAS LAST PRINTED

Secretary of Agriculture_____ HENRY A. WALLACE.
Under Secretary_____ M. L. WILSON.
Assistant Secretary_____ HARRY L. BROWN.
Coordinator of Land Use Planning and Director of Information_____ M. S. EISENHOWER.
Director of Extension Work_____ C. W. WARBURTON.
Director of Finance_____ W. A. JUMP.
Director of Personnel_____ ROY F. HENDRICKSON.
Director of Research_____ JAMES T. JARDINE.
Solicitor_____ MASTIN G. WHITE.
Agricultural Adjustment Administration_____ H. R. TOLLEY, Administrator.
Bureau of Agricultural Economics_____ A. G. BLACK, Chief.
Bureau of Agricultural Engineering_____ S. H. McCRORY, Chief.
Bureau of Animal Industry_____ JOHN R. MOHLER, Chief.
Bureau of Biological Survey_____ IRA N. GABRIELSON, Chief.
Bureau of Chemistry and Soils_____ HENRY G. KNIGHT, Chief.
Commodity Exchange Administration_____ J. W. T. DUVEL, Chief.
Bureau of Dairy Industry_____ O. E. REED, Chief.
Bureau of Entomology and Plant Quarantine__ LEE A. STRONG, Chief.
Office of Experiment Stations_____ JAMES T. JARDINE, Chief.
Farm Security Administration_____ W. W. ALEXANDER, Administrator.
Food and Drug Administration_____ WALTER G. CAMPBELL, Chief.
Forest Service_____ FERDINAND A. SILCOX, Chief.
Bureau of Home Economics_____ LOUISE STANLEY, Chief.
Library_____ CLARIBEL R. BARNETT, Librarian.
Bureau of Plant Industry_____ E. C. AUCHTER, Chief.
Bureau of Public Roads_____ THOMAS H. MACDONALD, Chief.
Soil Conservation Service_____ H. H. BENNETT, Chief.
Weather Bureau_____ C. C. CLARK, Acting Chief.

13

U. S. GOVERNMENT PRINTING OFFICE: 1938